Esdras Rocha da Silva
Ana Cecíia R. de Castro

Preservation and viability of cashew forks

Esdras Rocha da Silva
Ana Cecíia R. de Castro

Preservation and viability of cashew forks

Alternatives for germplasm banks using preservative solutions

ScienciaScripts

Imprint

Cover image: www.ingimage.com

This book is a translation from the original published under ISBN 978-3-330-76144-5.

Publisher:
Sciencia Scripts
is a trademark of
Dodo Books Indian Ocean Ltd. and OmniScriptum S.R.L publishing group

120 High Road, East Finchley, London, N2 9ED, United Kingdom
Str. Armeneasca 28/1, office 1, Chisinau MD-2012, Republic of Moldova, Europe
Managing Directors: Ieva Konstantinova, Victoria Ursu
info@omniscriptum.com

Printed at: see last page
ISBN: 978-620-8-37227-9

To my parents, for their infinite love, for guiding me along the paths of Truth and for everything they represent in my life.

Dedication.

ACKNOWLEDGEMENTS

To my Lord and Saviour Jesus Christ, sure Friend in uncertain gardens, Author and Finisher of my faith, for holding me firm in the palm of His hand, for giving me the ability to carry out this work and for revealing a little more of this mystical wisdom to His servant.

To my parents, Antonio Moreira da Silva and Ocioneide Rocha da Silva, for teaching me values such as integrity and honesty, for always being there for me in the most difficult situations and for the encouragement and motivation that drive me forward. You are the greatest blessing in my life.

To my siblings, Neubejamia, Neubejohnson and Neuila, who are always by my side, for their friendship, love, affection and for being part of my beloved family.

To EMBRAPA-CNPAT and Dr Ana Cecília, my supervisor at Embrapa, for her teachings, friendship, moments of relaxation and valuable help in my professional training.

To the UFC and to Professor Dr Márcio Cleber, for always being helpful and for his help in carrying out this work.

To Professor Maria Lúcia, the course coordinator, and Moisés, who have always been a support in the coordination office, helping me to solve administrative problems in an understanding and clear way.

To Dr Ana Cristina and Dr Diva, who always ensured the care and quality of the Embrapa - CNPAT Tissue Culture Laboratory.

To my colleagues and friends at EMBRAPA (Tissue Culture Laboratory and Genetic Resources and Plant Breeding Laboratory) and the UFC, who accompanied me during this long journey, for their friendship, help, companionship, jokes, moments of study and for making me laugh. In alphabetical order: Abelardo, Aldemir, Alexandre, Anderson, André Cardoso, André Rodrigues, Andrezão, Caiena, Cássia, Cíntia, Diel do "thrash", Éder, Érica, Evaldo, Gabi, Gislene, Heldo, lury, Jani, Joelma, Juliana Viana, Karina, Karlinha,

Kellina, Laíse, Márcio, Moacir, Otávio, Paula, Rafael Silvério, Raúl, Ravena, Sara, Silvana, Wanderson and Zirlane. Especially Anderson, Cayenne, Éder, Evaldo, Gislene and Iury.

To everyone I may have forgotten to put on this list, thank you.

"Bless the Lord, O my soul, and do not
squander any of his benefits."

Psalm 103:2

SUMMARY

Restoring degraded ecosystems is extremely important given the increasingly drastic environmental crisis and the decline in the quality of life of human and natural populations. The preservation of genetic resources in germplasm banks is an alternative for conserving the genetic variability present in a given species.

As a result of these factors, it is essential to create and use new viable techniques to maintain and conserve these plant genetic resources in the face of the unfeasibility of traditional seed collection techniques, since the high number of repetitions to represent a native population increases the cost of maintaining this bank, as well as the risk that this same population will not be fruiting at the time of shipment.

The use of cuttings from collected accessions ensures the accuracy of genetic patterns and reduces the number of repetitions to be planted in the bank. In Anacardium occidentale L., however, cuttings have low viability, and it is necessary to develop techniques that allow for greater conservation of these materials.

The aim of this study was to assess the conservation potential and viability of Anacardium occidentale L. forks when subjected to different concentrations of preservative solutions at different packaging intervals between collection and grafting. The design was entirely randomised, in a 3x4 factorial scheme. The treatments were storage in newspaper moistened with a plastic bag with or without an antioxidant solution; without storage and with time intervals for grafting (10 and 15 days after collection) with 7 replications and 2 control groups (cuttings without treatments and cuttings grafted on the day of collection). The plants were assessed at 60 days after grafting for survival, number of buds, number of leaf primordia and length of the grafted branch. The use of different concentrations of antioxidant solutions affects the setting of cuttings stored for 10 and 15 days after collection. The 1500 mg/L ascorbic acid solution is the one that best affects the sprouting of buds and the emission of leaf primordia in cuttings stored for 10 days. The 3 solutions evaluated affect bud sprouting, the emission of leaf primordia and the length of the grafted branch when stored for 15 days. However, it is necessary to develop other techniques that provide greater efficiency in the conservation and viability of conditioned cuttings.

Keywords: Anacardium occidentale L., genetic resources, germplasm.

SUMMARY

CHAPTER 1

INTRODUCTION

1.1 Literature Review

1.1.1 Economic Importance of Tropical Fruit Trees

In recent years, Brazilian fruit growing has become an increasingly important part of national and international agribusiness and has ample scope for expansion, given that the country has extensive areas with favourable climatic conditions for both temperate and tropical fruit trees.

More than a dozen species of fruit produced in different regions of Brazil (melon, grape, mango, apple, lemon, banana, papaya, watermelon, orange, cashew, pineapple, fig, avocado, coconut, mandarin, strawberry, cherry, etc) supply the planet, where more than 30 growing centres have been established, each exploiting its regional potential, in a mosaic of orchards that generates hundreds of thousands of jobs, drives the economy and brings the states into line with the reality of international trade (CARVALHO et al., 2010).

The agricultural base of the fruit production chain covers 2.2 million hectares (SEBRAE, 2015) and, according to Andrade (2010), generates 6.0 million direct jobs.

On the foreign market, the country has a good supply of tropical and temperate fruits for most of the year, thanks to its territorial extension, geographical position and favourable climate and soil conditions (ANDRADE, 2010).

In 2013, Brazilian fruit production was estimated at around 43.6 million tonnes (SEBRAE, 2015), indicating an increase on previous years since Brazilian fruit growing was not immune to the global economic crisis that spread in the second half of 2008, where production of 41.3 million tonnes in 2009 was down 4.0% on the previous year, when the volumes harvested were 43.0 million tonnes (CARVALHO et al., 2010).

Brazil is currently third in the world ranking of the largest fruit producers, behind only China and India, which produce 137.06 and 71.07 million tonnes of fruit respectively

(SEBRAE, 2015).

Brazilian fruit production grew by 19% in eight years (from 2001 to 2009), which is a favourable indicator for the expansion of fruit growing in the country. Consumption also increased, from 113kg per inhabitant per year to 125kg in the same period (FRUTICULTURA, 2011), with variations in subsequent years. Currently, however, due to the drought and the scenario of recession in the Brazilian economy between 2015 and 2016, it is possible that the country will see an average growth in sales of 1% to 1.5% by 2018 (TREICHEL et *al.,* 2016). The wide variety of fruit produced in all regions of Brazil comes from permanent and temporary crops, boosting opportunities for small businesses (SEBRAE, 2015) (Figure 1).

Figure 1 - Main producing states and fruit produced in Brazil. Source: SEBRAE, 2015.

Brazil's Northeast is a leading producer in the country and is one of the main vectors for economic development in semi-arid regions, as well as an important source of foreign currency. It is a pioneer in the production of bananas, coconuts, cocoa, cashews, papaya, pineapple, melons and passion fruit, and second in the production of grapes, oranges, lemons and guavas (ALMEIDA, 2008).

In 2007, Northeastern fruit production totalled 26% of total Brazilian production. With regard to exports, this figure jumped to 63 per cent in 2007. The states with the highest production in the Northeast are Bahia, Pernambuco, Ceará and Rio Grande do Norte. The main destination for Brazilian fruit exports is the European Union, with imports totalling around 70% of total national fruit exports. However, Brazilian fruit exports now reach 69 countries (CARVALHO, 2009).

1.1.2 The importance of cashew farming

The cashew tree occupies a prominent place among tropical fruit plants due to the growing commercialisation of its main products, such as the cashew nut kernel (ACC) and the liquid contained in the mesocarp of the nut, popularly known as cashew nut liquid (LCC); as well as other secondary products such as the floral stalk (false fruit), which is widely used in the fresh fruit market and for juices, sweets, soft drinks, etc.

The cashew agribusiness generates employment and income for thousands of people and foreign exchange for producing and exporting countries, almost all of which depend on the agricultural business in their economies (WALTER; CAVALCANTI, 2005).

CSA has earned a place in this market, as it is the most commercialised cashew product worldwide. Around the world, 240,000 tonnes/year of ACC are sold, which in 2006 resulted in around US$ 2.8 billion for the retail market. Recent studies suggest that ACC is rich in unsaturated fatty acids, i.e. it contains high levels of omega 9 and omega 6, which are essential throughout the human life cycle and could therefore increase its market value (GAZZOLA et al., 2007).

In the 2007 harvest, the area harvested and production in the main producing countries were estimated at 3,817,041 ha and 3,186,039 tonnes of nuts respectively, with Vietnam, Nigeria, India, Brazil, Indonesia, the Philippines, Ivory Coast, Tanzania and Guinea Bissau being the main cashew nut producers (TABLE 1). It is important to emphasise that Vietnam's leadership is due to an incentive programme with the planting of the early dwarf cashew tree (FIGUEIREDO JUNIOR, 2006), which originated in Brazil. In addition, it is a recent crop in the country and therefore has a lower incidence of phytosanitary problems. Another favourable factor is cheap labour, which in turn reduces costs. In Brazil, 176,000 tonnes of nuts were harvested in the 2007 harvest, resulting in exports of 41,569 tonnes of ACC, or 17% of world exports.

Table 1 - World cashew nut production (in tonnes) per year between 2001 and 2007.

Countries	Years						
	2001	2002	2003	2004	2005	2006	2007
Vietnam	292.80	515.20	657.600	818.800	960.800	941.600	961.00

	0	0					0
	485.00	514.00		555.00			660.00
Nigeria	0	0	524.0000		594.000	636.0000	
India	450.00	470.00	500.000	535.00	544.000	573.000	620.00
	0	0		0			0
	124.07	164.53					176.38
Brazil	3	9	183.094	187.839	152.751	243.7704	
	91.58	110.23					146.00
Indonesia	6	2	106.931	131.020	135.070	140.5730	
						113.07	118.00
Philippines	7.000	7.000	7.000	116.910	116.5331		0
Ivory Coast	87.57	104.98	84.811	140.636	175.000	200.000	130.00
	3	5					0
Tanzania	98.60	67.30	95.000	79.000	72.000	90.400	92.00
	0	0					0
Guinea-Bissau	85.00	86.00	86.000	86.000	88.000	81.000	81.00
	0	0					0
Mozambique	58.00	58.00	58.000	58.000	58.000	58.000	58.00
	0	0					0
Others	128.08	127.53	128.227	133.051	139.416	140.935	143.65
	8	5					5
	1.907.72	2.224.79	2.430.66	2.841.25	3.035.57	3.218.34	3.186.03
World	0	1	3	6	0	9	9

Source: FAO (2009)

In the north-east of Brazil, cashew farming is characterised by annual revenues of more than US$160 million. The activity generates more than 16,000 direct jobs in urban areas and more than 300,000 men/day/year in rural areas, the vast majority during the harvest period, which is equivalent to around 42,000 jobs a year (PAIVA et al., 2003).

The Brazilian cashew nut harvest for 2011, based on February 2011, showed a variation of +179.25 in production compared to 2010, -0.37% in planted area, +0.16% in harvested area (hectares) and +179.29% in yield (kg of nuts per hectare) (ANNEX), indicating the potential for expansion in this market, as well as gains in productivity.

Worldwide consumption continues to grow, indicating that there is a need for an increase in the production system, with the incorporation of more productive clones being the least costly factor given the generalised increase observed worldwide in the cost of agricultural inputs as a whole (CAVALCANTI; BARROS, 2009).

A report by the State Agricultural Research Centre of the Brazilian Institute of Geography and Statistics (IBGE) points to an expansion in the production and cultivation of early dwarf cashew trees in the state of Ceará. In 2008, the early dwarf cashew tree accounted for 19% of the state's cashew production, with almost 23,000 tonnes, and 11% of the area harvested with cashew trees in the state, with around 43,000 hectares. According to the report, the impact of the dwarf cashew tree on cashew productivity in Ceará was 10 per cent in 2008. Productivity was 312 kg/ha. Without the dwarf cashew tree, it is estimated that productivity would have been 285 kg/ha. Today, early dwarf cashew trees are grown in the states of Ceará, Piauí, Rio Grande do Norte, Maranhão, Pernambuco, Bahia, Pará, Tocantins, Mato Grosso and São Paulo, and are known worldwide for their precocity and high productivity in the case of some clones (PRODUÇÃO, 2010). Currently, however, Ceará stands out as Brazil's largest cashew producer, with an estimated production of 52,118 tonnes (ALMEIDA et a., 2017).

1.1.3 Genetic resources

During the 1970s, 1980s and 1990s, a number of scientists tried to define the term "plant genetic resources". These include Hawkes, (1997); Frankel and Bennett, (1970) and Brockhaus and Oetmann, (1996). There was no consensus and so a definition that adds value to Plant Genetic Resources was given by the FAO (1989).

According to this definition, plant genetic resources refer to the economic, scientific or social value of the hereditary material contained within and between species. They include, on the one hand, materials used in cytogenetics, evolutionary, physiological, biochemical, pathological or ecological research materials, as well as accessions evaluated for their agronomic or reproductive propensities. The diversity within these plants is called Phytogenetic Resources.

The Convention on Biological Diversity (CBD), on the other hand, defines this in Article 2:

> "Genetic material" means all material of plant, animal, microbial or other origin that contains functional units of heredity.
> [...]
> "Biological resources" means genetic resources, organisms or parts thereof, populations, or any other biotic component of ecosystems, of actual or potential utility or value to humanity.
> [...]
> "Genetic resources" means genetic material of actual or potential value

(BRASIL, 1998).

Plant genetic resources continue to play an important role in the development of agriculture. The world population is expected to increase by 2.6 billion in the coming years, from 6.5 billion today to 9.1 billion in 2050 (HAMMER; TEKLU, 2008). The world needs an astonishing increase in food production to feed this population. As such, plant genetic resources are the foundation on which agriculture and world food stocks are based and genetic diversity in germplasm collections is fundamental to the world's fight against hunger, constituting the raw material for multiplying new plant varieties and therefore serving as a reservoir of genetic diversity (CUQUMA, 2010).

1.1.1.1 Fruit genetic resources in Brazil

Brazil, due to its continental dimensions, has an immense floristic diversity, which is distributed across its different ecosystems and biomes (MARIANTE, 2009).

Among the existing categories, fruit species stand out for their high economic value, both in the fresh fruit trade and in the production of raw materials for agro-industry. In addition, many of these fruits are important sources of food and sustenance for low-income populations in various parts of the country. Despite all the importance of tropical fruit trees and their economic potential, many materials that are found in the wild or undomesticated state have a strong tendency to disappear due to the irrational exploitation of the ecosystems in which they occur (SILVA JUNIOR et al., 1999).

1.1.1.2 Cashew genetic resources

Cashew, Anacardium occidentale L, belongs to the Anacardiaceae family, which comprises around 60 to 74 genera and 400 to 600 species (RENDLE, 1938; BAILEY, 1964; KHOSLA et al., 1973; BRIZICKY, 1962; MITCHELL; MORI, 1987). The genus consists of trees and shrubs (rarely subshrubs and climbers), predominantly tropical and subtropical, with few representatives from temperate climates (WALTER; CAVALCANTI, 2005).

Anacardium occidentale L. is the most widely dispersed and only cultivated species in the genus, and is found throughout the tropical world. The natural distribution of this species, however, can be confused by dispersal by cultivation since, while the main centre of diversity of the genus is the Amazon region, with a secondary centre of diversity in the cerrados, the cashew tree can be found in various ecosystems in the north and northeast of Brazil (BARROS et.al., 1993). According to Paiva et al. (2003), most of the species' cultivated diversity is concentrated in north-eastern Brazil, causing some confusion between natural distribution and dispersal by cultivation.

According to Bailey (1942), the systematic position of the genus Anacardium is:

Division: Spermatophyta

Subdivision: Angiospermae

Class: Dicotyledoneae

Subclass: Archichlamideae

Order: Sapindales

Family: Anacardiaceae

Genre: Anacardium

It is also important to emphasise that, although the cashew tree is found more abundantly in the coastal region of the north-east of Brazil, where it grows in an apparently spontaneous state, especially in the vegetation of the beaches and dunes, it cannot withstand competition from other species, unlike other species of the genus, such as those of the cerrado and the Amazon rainforest, which normally coexist as part of the local flowers. Nor is it found spontaneously in the forests of the transition zones with the coastal strip in the north-east region (BARROS, 1993).

The systematic planting of certain genotypes has caused the genetic base of species to become increasingly narrow, leading to an ever greater standardisation of agricultural crops (COELHO; BARROS, 2005).

According to Coelho and Barros (2005), activities to collect and conserve wild species of the genus are practically non-existent, well below the demand that cashew farming needs in order to develop genotypes that provide greater productivity, as well as resistance and/or adaptation to biotic and abiotic stresses when under cultivation.

1.1.4 Germplasm conservation

The collection and introduction of materials, conservation and exchange, as well as the characterisation and evaluation of germplasm are all necessary and essential steps in the maintenance and use of genetic resources. Conservation provides support for genetic improvement work, enables the exchange of germplasm and, especially, the preservation of genetic variability, while characterisation and evaluation make it possible to know the quality and potential of germplasm in various respects (OLIVEIRA; SOARES FILHO, 1999).

The concentration of efforts in the genetic conservation of fruit species should be determined by the relative magnitude of genetic variation between and within populations, in order to collect and preserve the maximum genetic variability of the species. The loss of genetic variability caused by human activity is significant and is mainly due to the destruction of natural habitats of plant populations. This emphasises the importance of research and procedures aimed at conserving genetic resources in the tropical ecosystem (PAIVA et al., 2003).

The cashew tree, a predominantly allogamous species with a high degree of heterozygosity, requires large samples to represent the variability contained in natural populations. For this reason, germplasm conservation has a high cost, but it is highly important for the species' genetic improvement programme (PAIVA et al., 2003).

The traditional scheme for conserving plant genetic resources, which includes collection, conservation, characterisation, evaluation and use, has not usually been carried out efficiently in all these phases for tree fruit species, with the exception of those with high economic value and low genetic variability. Evaluation and characterisation are hampered as the number of accessions in the Active Germplasm Banks (AGBs) increases and human resources are not available to carry out these activities; consequently, the utilisation phase of the genetic variability collected is also hampered. Furthermore, both the collection and conservation of these resources are high-cost research activities with little economic return in the short term (PAIVA et al., 2003).

1.1.5 Cashew Active Germplasm Bank (BAG)

With the aim of collecting, conserving, characterising and evaluating the significant genetic variability available, the Active Cashew Germplasm Bank (BAG) of Embrapa Agroindústria Tropical was set up in the municipality of PACAJUS-CE.

Embrapa Agroindústria Tropical holds a cashew germplasm collection containing 588 accessions, the majority of which are of the A. occidentale L. species, and a minority of which are of other species: A. microcarpum Duck, A. othonianum Rizz, A. humile and accessions of Anacardium that have not yet been identified holding, according to Paiva et al. (2003), the greatest variability of accessions (FIGURE 1).

Figure 2 - Variability in the shapes and colours of cashew peduncles found in the active cashew germplasm bank at Embrapa - Agroindústria Tropical. Source: SOBREIRA JÚNIOR, 2010.

The states of origin of the accessions in the bank include: Ceará, Pará, Mato Grosso, Bahia, São Paulo, Piauí, Rio Grande do Norte, Roraima, Goiás, and there are also accessions from India and Venezuela (PAIVA et al., 2003).

Also according to Paiva et al. (2003), most of the accessions in the collection come from the state of Ceará, around 70 per cent, which deserves attention so as not to jeopardise the representativeness of the germplasm in the future. According to Coelho and Barros (2005), the vast majority of the accessions collected and conserved not only come from the state of

Ceará, but are also from the already cultivated species (A. occidentale L.), limiting the representativeness available in the collection.

The genetic improvement of cashew has its historical milestone when early dwarf cashew plants were introduced in that season, originating from a natural population in the municipality of Maranguape, CE (PAIVA et al., 2003); among the characters improved through the use of the collection, the following stand out: increases in productivity, peduncle quality and kernel weight; resistance to diseases; reduction in plant size; greater precocity and lengthening of the fruiting period (COELHO; BARROS, 2005).

The early dwarf cashew tree, also known as the six-month cashew tree, is characterised by its low size, homogeneous crown, stem diameter and wingspan that are much smaller than those of the common type. It starts flowering between six and 18 months. The weight of the fruit in natural populations varies from three to ten grams and of the peduncle from 20 to 160 grams. To date, individual production capacity has recorded production of up to 43kg of nuts (BARROS, 1988).

The genetic variability contained in the BAG-Cajueiro plant collection has made it possible to obtain early dwarf cashew clones recommended for commercial planting in the Northeast Region, from the early 1980s to the present day, such as CCP 06, CCP 09, CCP 76, CCP 1001, Epace CL-49, Embrapa 50, Embrapa 51, BRS 189 and BRS 226 (ALMEIDA et al., 1993; BARROS et al., 1984; BARROS, 1988; PAIVA et al., 2001), as well as 132 different clones of dwarf cashew and 40 of the common type still being evaluated in rainfed areas and under irrigation (PAIVA et al., 2003).

The collection is kept in the field at the Experimental Station in Pacajus-CE, with geographical coordinates of 4° 10' 21" S, 38° 27' 38" W and an altitude of 70 metres above sea level (IPECE, 2009), occupying a total area of 202 ha (COELHO; BARROS, 2005). The region has a hot tropical, mild semi-arid and hot sub-humid tropical climate, with average rainfall of 791.4 mm per year and average temperatures of 26° to 28°C (IPECE, 2009).

In addition to the accessions catalogued in the BAG, the Unit also has a working collection, used by the breeders, whose dynamics mean that various genotypes are constantly used and/or discarded, without detriment to the representativeness of the BAG (PAIVA et al., 2003).

1.1.6 Germplasm collection

The concentration of efforts in the genetic conservation of fruit species should be determined by the relative magnitude of genetic variation between and within populations, in order to collect and preserve the maximum genetic variability of the species. Losses of genetic variability caused by human activity are significant and are mainly due to the destruction of natural habitats of plant populations. This emphasises the importance of research and procedures aimed at conserving genetic resources in the tropical ecosystem (CAVALCANTI et al., 1997).

Quantified genetic diversity must be available to breeders if breeding programmes are to succeed in their task of producing clones adapted to the new environments in which the cashew tree is tested, which can be done through a comprehensive genetic resources programme with well-defined objectives (PAIVA et al., 2003).

Cashew germplasm is collected in two ways, in terms of the type of material collected. When the material collected is part of the plant for asexual multiplication, the aim is to clone and test the new clones. When the material collected is seeds, it is necessary to carry out a soft selection and then preserve it in collections kept in the field (PAIVA et al., 2003).

1.1.7 Grafting

The use of asexual reproduction techniques can make it possible to maintain and multiply a genotype with great fidelity to its genetic patterns (WALTER; CAVALCANTI, 2005), and grafting is one of the most widely used techniques in fruit propagation.

Grafting is the association between two parts of different plants that continue to grow as a single being. Two plants are considered: the scion or rootstock, which is the plant that contributes to the root system, ensuring mineral nutrition; and the scion or graft, which is the plant with noble characteristics that you want to reproduce, which forms the canopy and bears fruit, being responsible for absorbing sunlight and carbon from the air to transform raw sap into elaborate sap, essential to the life of the plant. Rootstocks are generally obtained from seeds ("free-rootstocks") and grafts are obtained from the branches of the selected mother plants that are to be propagated (RIBEIRO et al., 2005).

Among the aspects that recommend grafting are: early production (transfer of maturity), reduction in plant size (facilitates cultivation), making it possible to grow species or varieties susceptible to phytosanitary and/or environmental problems, maintenance/expansion of desirable characteristics segregated by natural or induced mutations, preservation/multiplication of noble varieties (in terms of quality and productivity), avoidance of undesirable segregations, renewal of declining orchards, replacement of uninteresting plants; restoration of damaged plants, ornamental or exotic character (floriculture, or multivarietal plant) and studies or indexing tests for viruses (exocort, sorosis, xyloporosis) (RIBEIRO et al., 2005).

However, several factors make grafting incompatible: the lack of anatomical affinity in terms of cell size and formation, the nutritional requirements and life cycle of the tissues involved, the size and vigour of the scion and rootstock cultivars, the consistency of the tissues, the oxidation of phenolic substances which hinders the formation of callus, and the presence of phytopathogenic microorganisms in both parts (FACHINELLO et al., 1995).

The removal of forks for grafting during the reproductive growth phases (flowering and fruiting) is not recommended because at this time the plant allocates its reserves to the formation of the reproductive organs, leaving the cambial tissue with low availability of carbohydrates to heal the wound caused by the operation. As well as aspects related to cellular activity, there is also the plant's hormonal situation, which changes during the different phenological phases and may or may not favour graft healing (HARTMANN et al., 2002).

Corrêa et al. (1993) commented that among the methods of vegetative propagation, grafting is currently the most widely used method for cashew propagation.

Several methods of vegetative propagation can be used to form cashew seedlings. However, because they are more technically and economically viable, only two are used: grafting by bubbling on a plate and grafting by grafting, which can be either lateral grafting or full slit grafting, also known as top grafting (CAVALCANTI JÚNIOR; CHAVES, 2001).

Cashew forks, however, are only viable for grafting for a short period of time, so collections made using vegetative propagules require a short time to return to the nursery where the grafts will be made. This is a limiting factor in the case of longer collections, which forces the preference for collecting by seed (SOBREIRA JÚNIOR, 2010).

Due to the low durability of forks, it is necessary to develop methodologies for preserving forks in order to increase their viability during the period between transport and seedling preparation, thus facilitating the management of BAG activities.

Just as important as the collection and transport of vegetative material is the proper packaging of this material, which is fundamental for successful propagation. The scarcity of material in the literature on collecting material for germplasm and on packaging techniques for cashew cuttings indicates a potential area of study that needs to be explored by the scientific community.

1.1.8 Antioxidant solutions (preservatives)

A broad definition that can be given to compounds with antioxidant activity is the following: that substance which, present in low concentrations compared to the oxidisable substrate, effectively delays or even inhibits the oxidation of that substrate (SIES; STAHL, 1995). In the food industry, antioxidants have the function of preventing product deterioration and maintaining nutritional value. They are also of great interest to biochemists and health professionals, helping to protect the body against damage caused by reactive oxygen species and regenerative diseases (SHAHIDI, 1996). According to Araújo (2008), the prevention of oxidation in plant tissues can be carried out by:

> thermal inactivation of the enzyme through the use of heat;
> *and* exclusion or removal of one or both substrates (oxygen, enzyme and substrate);
> reduction in pH by two or more units below the optimum pH (6.0);
> the addition of reducing substances that inhibit the action of polyphenol oxidase (PPO) or prevent the formation of melanin (such as ascorbic acid or citric acid).

Ascorbic acid (FIGURE 2) is recognised for its reducing action and nutritional contribution (vitamin C). This acid and its various neutral salts are the main antioxidants for use in fruit and vegetables and their juices to prevent browning and other oxidative reactions (WILEY, 1994). Citric acid (FIGURE 2) is one of the main natural organic acids in fruit, preventing enzymatic browning by acting on PPOs and peroxidases (enzymes responsible for enzymatic browning in fruit and vegetables). It is also used to potentiate (synergise) the action of other substances such as ascorbic acid itself (CHITARRA, 2000). Both ascorbic acid and citric acid are natural antioxidant compounds and have the ability to reduce the quinones formed by the action of oxidases, thus preventing the formation of blackened products, as well as acting as inhibitors of oxidative enzymes by lowering the pH (BEZERRA et al., 2002).

Cysteine (FIGURE 2) is an amino acid that has been successfully used to preserve bananas (MELO; VILAS-BOAS, 2006), apples and potatoes (MOLNAR-PEARL; FRIEDMAN, 1990; ROCCULI et al., 2007). Three different mechanisms of action of cysteine are proposed: reduction of o-quinones to o-dihydroxyphenols (KAHN, 1985); direct inhibition of polyphenol oxidase activity (DUDLEY; HOTCHKISS, 1989) and reaction with o-quinones giving rise to colourless cysteine-quinone compounds, competitive inhibitors of polyphenol oxidase (RICHARD-FORGET et al., 1992).

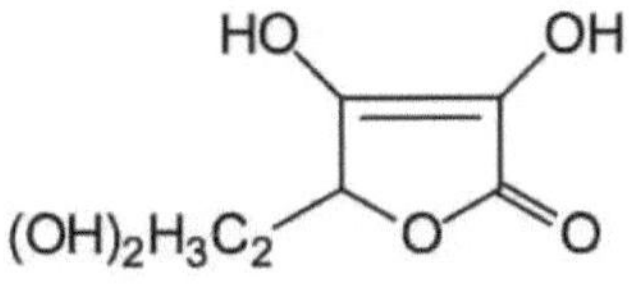

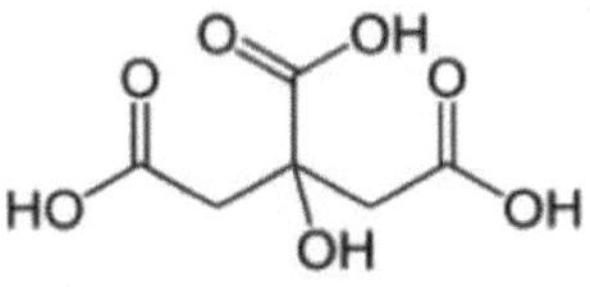

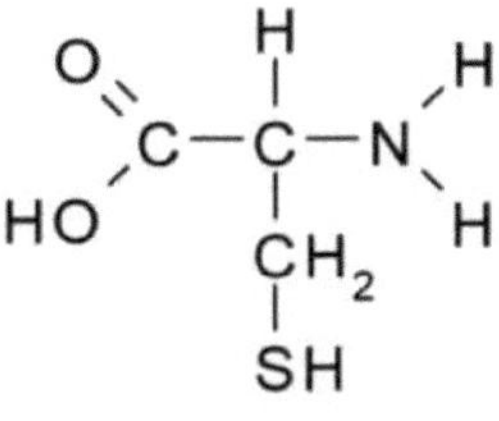

Figure 3 - Chemical structure of some substances used as antioxidant compounds.

Little is known, however, about the action of these natural antioxidant compounds and their effects on the preservation and viability of fruit forks, and research is needed to formulate methodologies and techniques that can extend the shelf life of these propagules.

CHAPTER 2

BACKGROUND

Restoring degraded ecosystems is extremely important given the increasingly drastic environmental crisis and the decline in the quality of life of human and natural populations.

Degraded ecosystems generally have a decrease in diversity and suffer successive disturbances that lead to a reduction in resilience and loss of stability. Preserving genetic resources in germplasm banks is an alternative for conserving the genetic variability present in a given species.

As a result of these factors, it is essential to create and use new viable techniques to maintain and conserve these plant genetic resources, given the unfeasibility of traditional seed collection techniques due to the high number of repetitions to represent a native population, which consequently increases the cost of maintaining this bank, as well as the high risk that this population will not be fruiting at the time of the expedition.

The use of cuttings has emerged as an alternative for the maintenance and conservation of plant genetic material, due to their genetic fidelity. The development of methodologies to enable this conservation is extremely important due to the low viability of these propagules.

CHAPTER 3

OBJECTIVE

To assess the preservation potential and viability of Anacardium occidentale L. forks when subjected to different concentrations of preservative solutions at different packaging intervals between collection and grafting.

CHAPTER 4

MATERIAL AND METHODS

All stages of the experiment were carried out in the Experimental Field of Embrapa Agroindústria Tropical, in the municipality of Pacajus-CE, with geographical coordinates of 4° 10' 21" S, 38° 27' 38" W and an altitude of 70 metres above sea level (COELHO; BARROS, 2005). The region has a hot tropical, mild semi-arid and hot sub-humid tropical climate, with average rainfall of 791.4 mm per year and average temperatures of 26° to 28° C (IPECE, 2009).

The study was set up in a completely randomised design, with a 3x4 factorial scheme (storage x preservative respectively), where the storage factor was the interval during which the cuttings were stored (from collection to grafting), in the following periods: No Storage (SA), 10 days of storage (A1) and 15 days of storage (A2) and as a preservative factor, cuttings treated with Preservative Solution 1 (SC1): 150 mg/L ascorbic acid + 25 mg/L citric acid + 2 mg/L L-cysteine, obtained from Zaccaro et al. (2006); Preservative Solution 2 (SC2): 150 mg/L ascorbic acid, Preservative Solution 3 (SC3): 1500 mg/L ascorbic acid and cuttings without preservative solution (SSC), totalling 9 treatments, two control groups and one absolute control group (TABLE 2).

Table 2 - Description of all the treatments and control groups used to set up the experiment.

Treatment	Description
T1	Cuttings immersed in Solution 1 and grafted on the day of collection
T2	Cuttings immersed in Solution 2 and grafted on the day of collection
T3	Cuttings immersed in Solution 3 and grafted on the day of collection
T4	Cuttings immersed in Solution 1 and conditioned, grafted 10 days after collection
T5	Cuttings immersed in Solution 2 and conditioned, grafted 10 days after collection
T6	Cuttings immersed in Solution 3 and conditioned, grafted 10 days after collection
T7	Cuttings immersed in Solution 1 and conditioned, grafted 15 days after collection
T8	Cuttings immersed in Solution 2 and conditioned, grafted 15

	days after collection
T9	Cuttings immersed in Solution 3 and conditioned, grafted 15 days after collection
Control 1 (C1)	Cuttings untreated and conditioned after collection, grafted 10 days after collection
Control 2 (C2)	Cuttings untreated and conditioned after collection, grafted 15 days after collection
Absolute Control (AC)	Cuttings collected and grafted in the first collection

4.1 Material collection

A. occidentale L. forks, 10.5 cm long, were collected from the tip branches of the mother plants of a clonal garden, clone CCP 76 (FIGURE 3), which is used in the commercial production of seedlings, then the leaf petioles were removed from these tips. A total of 84 cuttings were used, collected on the same day by the same collector.

Figure 4 - Collecting cuttings in the clonal garden.

4.2 Assembly, Simulation and Grafting of Treatments

The cuttings were grouped into seven per elastic cord until the moment of grafting and dipped in a preservative solution according to Table 2, where they remained for 2 minutes, with the exception of the control treatments (CA, C1 and C2) which received no preservative solution.

Treatments T1, T2, T3 and CA (FIGURE 4) were grafted on the day of collection. The other cuttings were packed in newspaper moistened with distilled water inside PVC plastic bags, sealed with durex tape, placed in standardised polystyrene boxes 22 X 29 X 20 cm in size, covered and stored at room temperature and grafted 10 days after collection (T4, T5, T6 and C1) and 15 days after collection (T7, T8, T9 and C2) (FIGURE 5).

After completing the simulation periods corresponding to 10 and 15 days after collection, the respective cuttings were immediately grafted. The rootstocks used were seedlings approximately 30 cm long and 90 days old, obtained from seeds, also of the CCP 76 clone, sown in rigid polypropylene tubes 190 mm high, 52 mm in diameter and with a capacity of 288 cm .^{3}

Figure 5: Freshly collected cashew cuttings that have not been conditioned.

Figure 6 - Storage of treatments (2 in each box).

The grafting method used was lateral slit grafting. The procedure begins with the extraction of the rootstock's apical meristem, after which a small cut is made across the length of the stem (lateral slit), at a height of 15 cm from the plant's root, until the vascular cambium is exposed. After this, a small fragment of the end of the cutting is cut off so that it fits into the rootstock slit, bringing the cambium of both into direct contact.

The parts were joined together by tying vinyl plastic around the crack area, so that the cut was not exposed to drying out or contamination (FIGURE 6).

Figure 7 - Detail of the tie made with vinyl plastic around the junction of the scion and rootstock in cashew seedlings.

All the procedures described were carried out on all the seedlings by the same grafter, given the need to standardise the experiment as much as possible.

The existing leaves on the rootstock were kept above the grafting point, totalling 4 leaves, in an effort to maintain the plant's photosynthetic activity until the rootstock and fork tissues joined.

After grafting, the cuttings were completely covered with a cellophane-type plastic bag (4 x 20 cm) to form a humid chamber to prevent the tissues from dehydrating (FIGURE 7). The bags were removed as soon as the cuttings showed expanded leaf primordia.

An assessment was made at 60 DAE (days after grafting), a period when grafted seedlings intended for sale are usually ready for marketing and planting.

Figure 8 - Grafted cashew cuttings wrapped in cellophane.

4.3 Experiment evaluation

After grafting, the seedlings were placed in a completely randomised design in a nursery at the Pacajus unit (Figure 8), approximately 2.5 m long and 50% shaded, where they remained on benches at a height of 36 cm above the ground under daily manual watering.

Figure 9 - Seedlings in the nursery after grafting.

The seedlings were analysed every fortnight from the date of grafting, taking into account the following aspects:

1. Pegging

   Analyses of the percentage of cuttings taken for each treatment and time interval of the simulation, where cuttings that showed obvious signs of necrosis in the tissues or that had not sprouted leaves by the end of the experiment were considered dead.

2. Number of buds

   Number of buds produced by the cutting.

3. Number of leaves

   Number of leaf primordia on the cutting.

4. Pile length

   Length of the grafted branch, from the scar to the stem apex, measured in centimetres using a ruler.

4.4 Statistical Data Analysis

Statistical analysis was carried out using the ESTAT programme developed by FCAV/UNESP, Jaboticabal, SP. Tukey's test ($\alpha=0.05$) was used to compare the means. The "setting" variable was calculated as the percentage of live cuttings at 60 DAE using the Microsoft Excel 2010 programme .©

CHAPTER 5

RESULTS AND DISCUSSION

After the storage intervals, differences were observed in the appearance and vigour of the cuttings as a function of the treatments used (FIGURE 9). This difference was reflected in a variation in the pegging indices of the cuttings, and measurement was inferred only in the cuttings that were successfully grafted (TABLE 3).

Treatments without a preservative solution or with a longer storage time until grafting suffered more browning than the other treatments. According to Fachinello et al. (1995), in some species, oxidation of phenolic compounds occurs where the cutting is made, and this oxidation is observed by the darkening of the tissue.

George (1984 apud DESCHAMPS, 1993) explains that oxidation of phenolic compounds occurs in damaged tissues due to the enzymatic action of oxidase, i.e. the ends of the tissue darken quickly and toxic oxidation products are released. Even with antioxidant treatment, it was possible to see that the length of storage time played a key role in whether or not the damaged tissues of the cuttings darkened. Visual differences were also observed in the turgidity of the packaged cuttings and, in some treatments, abscission of the pieces of leaf petioles remaining after the cuttings were prepared.

The treatments with the highest setting rates were those collected and grafted on the same day as collection (T1, T2, T3 and CA) with maximum setting rates in percentage (TABLE 3). According to Sobreira Júnior (2010), it is possible to obtain 100% setting rates in cashew cuttings collected and grafted immediately, without the use of antioxidant solutions at 60 DAE.

According to Peil (2003), for grafting to be successful, the forks must preferably be of the same vigour as the rootstock, since the healing of the grafting point depends on the production of parenchymatous cells in the cambium regions of both, which then mix and intertwine, forming what is normally known as "callus".

The subsequent treatments (T4, T5, T6, T7, T8, T9, C1 and C2) varied. There was a progressive decrease in the percentage of grafting as the storage time until grafting increased. In cuttings stored for 10 days, treatment T6 had the highest rate of setting with approximately 86% and T4 had the lowest rate for the same period, around 57%.

Table 3 - Pegging percentage in the treatments evaluated at 60 DAE, where: T1= S1+SA, T2= S2+SA, T3= S3+SA, T4= S1+A1, T5= S2+A1, T6= S3+A1, T7= S1+A2, T8= S2+A2, T9= S3+A2, CA= SSC+SA, C1= SSC+A1 and C2= SSC+A2.

Treatment	%
T1	100
T2	100
T3	100
T4	57,1
T5	57,1
T6	85,7
T7	28,6
T8	85,7
T9	42,9
CA	100
C1	28,6
C2	0

*S1= cuttings immersed in solution 1, S2= cuttings immersed in solution 2, S3= cuttings immersed in solution 3, SSC= cuttings without preservative solution, SA= cuttings without conditioning (grafting on the day of collection), A1= conditioned cuttings (grafting at 10 days of conditioning) and A2= conditioned cuttings (grafting at 15 days of conditioning).

In general, storing the cuttings for 15 days caused a significant reduction in the survival of the cuttings at 60 DAE, especially in the cuttings from the control treatment (C2), which were not exposed to antioxidant solutions, whose survival rate was 0%. However, treatment T8 showed a much higher value, with almost 86% of the cuttings alive during the period analysed.

The time of grafting is one of the external factors that can affect setting. Normally, tropical species show optimum setting rates when grafting is carried out at temperatures around 30°C, when there is greater cambial activity (HARTMANN et al., 2002). The entire process of the experiment (from the collection of forks to the evaluation at 60 DAE) was carried out between January and April 2011, when the average rainfall was 251.5 mm, according to the Ceará Foundation for Meteorology and Water Resources (FUNCEME,

2011). The average temperature for the region is 26 to 28° C (IPECE, 2009), which may have contributed to the greater stability of the grafted cuttings.

Humidity is also related to the good preservation of plant material and, consequently, to good setting results after grafting intervals. Very low humidity levels favour the rapid dehydration of tissues, causing wilting and nutritional losses. At the same time, high humidity levels can create an ideal environment for the development of fungal pathogen infections (BRACKMANN et al., 2005).

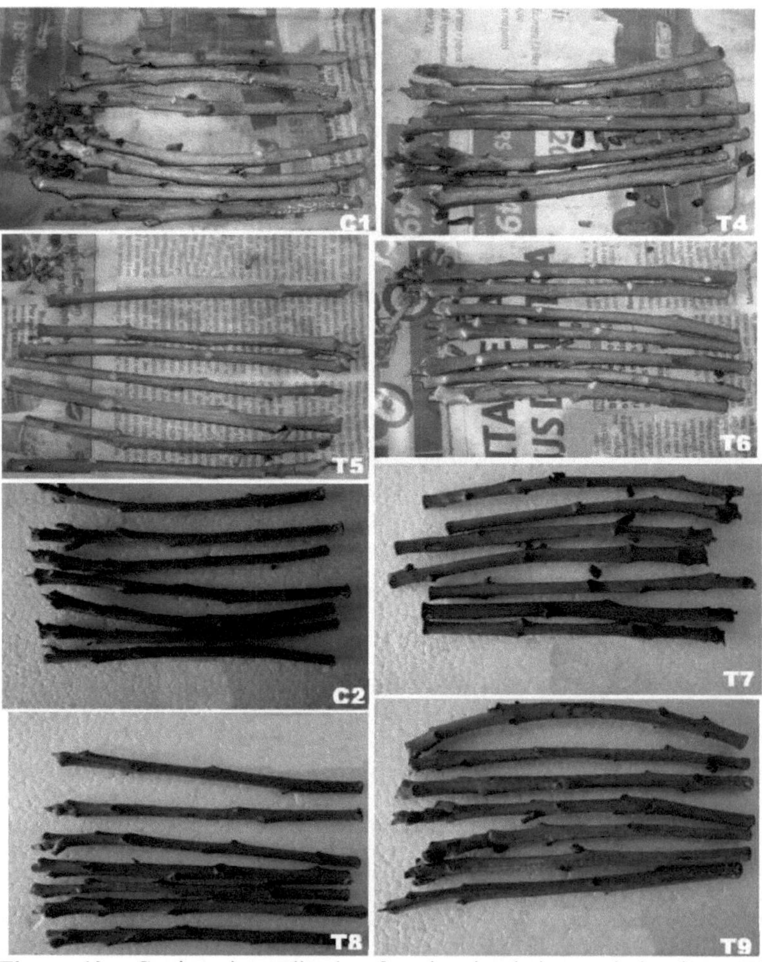

Figure 10 - Cuttings immediately after the simulation period, where: T1= S1+SA, T2= S2+SA, T3= S3+SA, T4= S1+A1, T5= S2+A1, T6= S3+A1, T7= S1+A2, T8= S2+A2, T9= S3+A2, C1= SSC+A1 and C2= SSC+A2.*S1= cuttings immersed in solution 1, S2= cuttings

immersed in solution 2, S3= cuttings immersed in solution 3, SSC= cuttings without preservative solution, SA= cuttings without conditioning (grafting on the day of collection), A1= conditioned cuttings (grafting at 10 days of conditioning) and A2= conditioned cuttings (grafting at 15 days of conditioning).

According to Goto (2003), low humidity inhibits the formation of callus, increasing cell desiccation and therefore high relative humidity is necessary to avoid tissue dehydration. Also according to Funceme (2011), the relative humidity found in the first interval of the experiment was 91.2%, which may have contributed to the success of the grafting in that period.

According to Zaccaro et al. (2006), with mango cultivars (Mangifera indica L.), despite having components that reduce oxidation rates in the materials, it seems that cuttings treated with an antioxidant solution did not have significantly higher setting rates than those not treated with the same solution.

According to Sobreira Júnior (2010), the proliferation of pathogens on the cuttings can interfere with the setting rates of the grafts. These agents accelerate the process of necrosis of meristematic tissues, jeopardising the development of the cutting after it is joined to the rootstock. However, no organic matter or fungicides were found in any of the cuttings used in the experiment. There were significant differences in the variables analysed at 60 DAE, according to the summary of the analysis of variance, using the F test (a=0.05) (TABLE 4).

Table 4 - Summary of the Analysis of Variance, of cuttings conditioned under different storage times and different preservative solutions, where: NG= Number of buds emitted, NPF= Number of leaf primordia emitted, CR= Branch length, SA= No storage, A1= 10-day storage, A2= 15-day storage, SC= No preservative, C1= With preservative 1, C2= With preservative 2 and C3= With preservative 3.

		MEAN SQUARES		
FV	GL	NG	NPF	CR
Storage (A)	2	0,1944*	3,6586 NS	49,7469**
Preservative (C)	3	0,5485**	33,4655**	66,3648**
A x C	6	0,4419**	7,2594*	42,1540**
Waste	72	0,0533	2,7694	0,6781
CV		22,2%	32,5%	6,40%
		DEVELOPMENTS		
FV	GL	NG	NPF	CR
C (SA)	3	0,0200 NS	0,7844 NS	1,9622 NS
C (A1)	3	0,2700**	19,8964 **	0,4142 NS
C (A2)	3	1,1422**	27,3033**	148,2964**

A (SSC)	2	1,1944**	16,3611**	174,8211**
A (SC1)	2	0,00 NS	2,2044 NS	0,1033 NS
A (SC2)	2	0,05444 NS	2,2800 NS	1,2411 NS
A (SC3)	2	0,2711 *	4,5911 NS	0,0433 NS

*Significant at 5% probability by Tukey's test; ** Significant at 1% probability by Tukey's test; NS Not significant

Storage of the cuttings affected the number of buds (NG), p=0.01, and branch length (CR), p=0.05. The number of leaf primordia (NPF) was not affected by storage. On the other hand, the preservative factor affected all the variables evaluated, p=0.01.

The interaction between the factors revealed p=0.01 for NG and CR, while p=0.05 for NPF. The breakdown of the analysis showed that the preservative factor without storage was not significant in any of the variables assessed, but at 10 and 15 days of storage, the preservative factor affected all the variables analysed (p=0.01), with the exception of CR at 10 days of storage.

The storage intervals used affected the treatments that did not involve the use of an antioxidant solution (CA, C1 and C2) in all the variables evaluated (p=0.01). However, no significant effect was observed for any of the preservative solutions used among the characters assessed, except within SC3 for NG (p=0.05).

The use of preservatives did not affect budbreak when grafting was carried out immediately after collection (TABLE 5). At the 10-day storage interval, SC3 differed from the other treatments, with an average of 1.6 buds emerging.

Table 5 - Number of buds, where: SA= No Storage, A1 = 10-day Storage, A2= 15-day Storage, SSC= No Preservative Solution, SC1= Preservative Solution 1, SC2= Preservative Solution 2 and SC3= Preservative Solution 3.

STORAGE		**CONSERVANTS**			**AVERAGE**
	SSC	SC1	SC2	SC3	
SA	1.17 Aa	1.00 Aa	1.10 Aa	1.00 Aa	1.07 ab
A1	1.00 Ba	1.00 Ba	1.00 Ba	1.60 Aa	1,15 a
A2	0.00 Bb	1.00 Aa	1.27 Aa	1.33 Aab	0,90 b
AVERAGE	o,	1,	1.12 AB	1,31 A	

Equal uppercase letters in the same row and equal lowercase letters in the same column do not differ statistically according to the Tukey test (P< 0.05).

At 15 days after collection and storage, the treatments exposed to antioxidant solutions were successful in producing buds, unlike the treatment that did not receive antioxidant solution, which was unsuccessful in grafting.

Cuttings that were not immersed in the antioxidant solution sprouted more buds when grafted immediately after collection. The treatments involving solution 1 (SC1) and solution 2 (SC2), when compared to each other, did not significantly affect the emission of buds in the time intervals evaluated, with the exception of solution 3 (SC3), where they did not differ statistically in terms of grafting on the day of collection and grafting after 10 days of storage. The treatment stored for 15 days using this solution obtained intermediate values compared to the other periods (p=0.05).With regard to the emission of leaf primordia, the use of a preservative did not affect the values.

grafted treatments immediately after collection (TABLE 6).

Table 6 - Number of leaf primordia, where: SA= No Storage, A1= 10-day Storage, A2= 15-day Storage, SSC= No Preservative Solution, SC1= Preservative Solution 1, SC2= Preservative Solution 2 and SC3= Preservative Solution 3.

STORAGE	**CONSERVANTS**				**AVERAGE**
	SSC	SC1	SC2	SC3	
SA	4,6 Aa 7	5,77Aa	5,73Aa	5,43Aa	5,4 a 0
A1	2,5 Bab 0	4,17ABa	7,33Aa	7,90Aa	5,4 a 8
A2	0,0 Bb 0	5,50Aa	5,93Aa	6,50Aa	4,4 a 8
AVERAGE	2,3 B 9	5,14A	6,33A	6,61A	

Equal uppercase letters in the same row and equal lowercase letters in the same column do not differ statistically according to the Tukey test (P< 0.05).

However, for the 10-day storage interval, the SC2 and SC3 solutions positively affected leaf emission, with an average of approximately 8 leaves using the SC3 solution. For a storage interval of 15 days, the use of the antioxidant solution was essential for the emission of leaves since the control treatment, for the same period, did not provide vegetative development of the cuttings.

Grafting with freshly collected cuttings, without antioxidant treatment, differed from grafting with packaging, with an average of almost 5 leaf primordia. At the same time, when comparing the storage factor individually within each treatment submitted to the preservative solution, no difference was observed. Cuttings stored for 15 days were not successfully grafted and consequently did not develop branches (TABLE 7). The use of antioxidant solutions affected the treatments that were stored for this period (p=0.01). In relation to the other periods, the use of the solutions evaluated did not affect the length of vegetative branches. Cuttings grafted without the use of antioxidant solutions promoted branch development when grafted immediately after collection or when stored for up to 10 days.

Table 7 - Branch length, where: SA= No Storage, A1= 10-day Storage, A2= 15-day Storage, SSC= No Preservative Solution, SC1= Preservative Solution 1, SC2= Preservative Solution 2 and SC3= Preservative Solution 3.

STORAGE	**CONSERVANTS**				**AVERAGE**
	SSC	SC1	SC2	SC3	
SA	12,60 Aa	14,13 Aa	14,37Aa	14,10 Aa	13,80 a
A1	13,77 Aa	14,27 Aa	14,67Aa	14,33 Aa	14,26 a
A2	0,00 Bb	14,50 Aa	13,43Aa	14,17 Aa	10,53 b
AVERAGE	8,79 B	14,30 A	14,16A	14,20 A	

Equal uppercase letters in the same row and equal lowercase letters in the same column do not differ statistically according to the Tukey test ($P< 0.05$).

Temperature is another factor that can affect the grafting of packaged cuttings, as well as their post-grafting development.

However, according to Seibert et al. (2008), low temperatures reduce the metabolism of plant materials, preventing rapid deterioration and oxidation, as well as minimising the reproduction of pathogens. However, prolonged storage at temperatures close to 2°C can lead to severe cold damage.

According to Sobreira Júnior (2010), when evaluating the conservation of cashew tree forks, treatments that involved the use of refrigeration compromised the survival of the cuttings, with mortality rates of around 60 per cent.

Figure 10 shows the situation of the cuttings evaluated at 60 DAE, which were grafted immediately after collection, and figure 11 shows the situation of the cuttings evaluated at 60 DAE, from storage 10 and 15 days after collection.

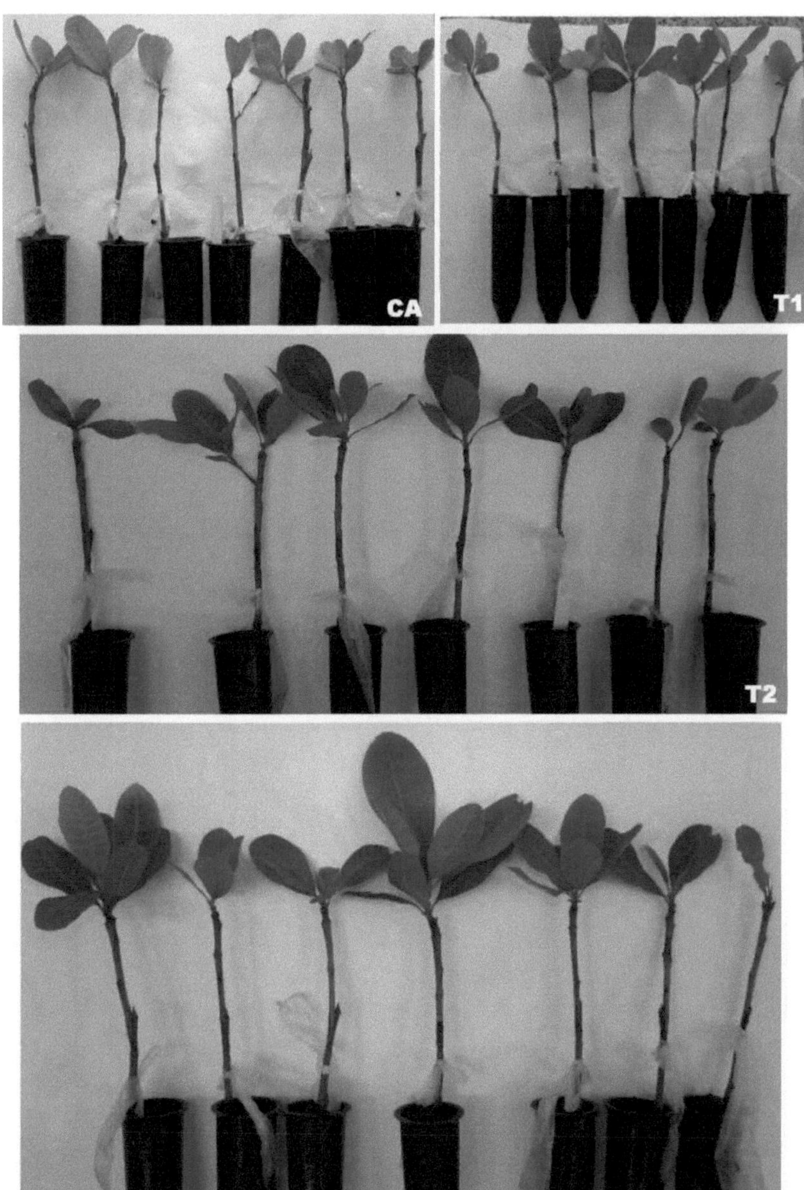

Figure 11 - Unconditioned cuttings, grafted on the day of collection, at 60 DAE, where: T1 = cuttings immersed in solution 1 (150 mg/L ascorbic acid + 25 mg/L citric acid + 2 mg/L L-cysteine), T2= cuttings immersed in solution 2 (150 mg/L ascorbic acid), T3= cuttings immersed in solution 3 (1500 mg/L ascorbic acid) and CA= cuttings without preservative solution.

Figure 12 - Conditioned and grafted cuttings after conditioning, at 60 DAE, where: T4= S1+A1, T5= S2+A1, T6= S3+A1, T7= S1+A2, T8= S2+A2, T9= S3+A2, C1= SSC+A1.*S1 = cuttings immersed in solution 1 (150 mg/L ascorbic acid + 25 mg/L citric acid + 2 mg/L L-cysteine), S2= cuttings immersed in solution 2 (150 mg/L ascorbic acid), S3= cuttings immersed in solution 3 (1500 mg/L ascorbic acid), SSC= cuttings without preservative solution, A1= grafting at 10 days of conditioning and A2= grafting at 15 days of conditioning.

CHAPTER 6

CONCLUSION

In general, the most efficient period for successful grafting, as well as for the viability of the cuttings, was when the propagules were not stored, regardless of the use of antioxidant solutions. The use of different concentrations of antioxidant solutions did not affect the variables evaluated when the cuttings were grafted on the same day as collection, but it did affect the setting of cuttings stored for 10 and 15 days after collection. Among the solutions evaluated, the 1500 mg/L ascorbic acid solution is the one that best affects bud sprouting and the emission of leaf primordia in cuttings stored for 10 days. The 3 solutions evaluated affect the setting, bud sprouting, emission of leaf primordia and the length of the grafted branch when stored for 15 days. However, it is necessary to develop other techniques that provide greater efficiency in the conservation and viability of conditioned cuttings.

CHAPTER 7

REFERENCES

ALMEIDA, C.V. de. Brazilian fruit growing analysed. **Portal do Agronegócio**, Jul. 2008. Available at: <http://www.portaldoagronegocio.com.br/conteudo.php?id=24830>. Accessed on: 15 Apr. 2011.

ALMEIDA, J.I.L.; ARAÚJO, F.E.; LOPES, J.G.V. **Evolution of the early dwarf cashew tree at the Pacajus experimental station, Ceará.** Fortaleza: EPACE, 1993. 17p. (EPACE. Documents, 6).

ALMEIDA, L.H.F.; CORDEIRO, S.A.; PEREIRA, R.S.; COUTO, L.C.; LACERDA, K.W.S. **Economic viability of cashew** (Anacardium occidentale **L.) production**. Nativa, Sinop, v.5, n.1, p.9-15, jan./feb. 2017.

ANDRADE, P.F.Z. Analysis of the agricultural situation - 2010/2011 harvest. **Secretariat of Agriculture and Supply** - Department of Rural Economy, PR, Nov. 2010. Available at: <http://www.seab.pr.gov.br/arquivos/File/deral/Prognosticos/fruticultura_2010_11.pdf>. Accessed on: 15 Apr. 2011.

ARAÚJO, J.M.A. **Química de alimentos**: teoria e prática. 4. ed. Viçosa, SC: Ed. Universidade Federal de Viçosa, 2008. 596p.

BAILEY, L.H. **Manual of cultivated plants**. 8. ed. New York: Macmillan, 1964. 1.116p.

BARROS, L. de M.; ARAÚJO, F.E.; ALMEIDA, J.I.L.; TEIXEIRA, L.M.S. **A cultura do cajueiro anão**. Fortaleza: EPACE, 1984. 67p. (EPACE. Documentos, 3).

BARROS, L. de M. Improvement. In: LIMA, V.P.M.S. **A cultura do cajueiro no nordeste do Brasil**. Fortaleza: BNB/ETENE, 1988. p. 321-355.

BARROS, L. de M.; PIMENTEL, C. R. M. P.; CORREA, M. P. F.; MESQUITA, A. L. M. **Technical recommendations for growing early dwarf cashew** trees. Fortaleza: EMBRAPA/CNPAT, 1993. 65p. (EMBRAPA-CNPAT. Technical Circular 1).

BEZERRA, V. S.; PEREIRA, R. G. F. A.; CARVALHO, V. D.; VILELA, E. R. Minimally processed cassava roots: effect of bleaching on quality and conservation. **Ciência Agrotécnica**, Lavras, v. 26, n. 3, p.564-575, 2002.

BRACKMANN, A.; PINTO, J. A. V.; STEFFENS, C. A.; GUARIENTI, A. J. W.; GIEHL, R. F. H.; SESTARI, I. Consequence of relative humidity during refrigerated and controlled atmosphere storage on the quality of 'Gala' apples. **Ciência Rural**, Santa Maria, v.35, n.5, p. 1197-1200, 2005.

BRAZIL. Decree No. 2.519, of 16 March 1998. Promulgates the Convention on Biological Diversity, signed in Rio de Janeiro on 5 June 1992. **Presidency of the Federative Republic of Brazil**. Available at: <https://www.planalto.gov.br/ccivil_03/decreto/D2519.htm>. Accessed on: 15 April 2011.

BRIZICKY, G.K. The genera of Anacardiaceae in the southeastern United States. **Journal of Arnold Arboretum**, v.43, p. 359-65, 1962.

BROCKHAUS, R. AND OETMANN, A. Aspects of the documentation of in situ conservation measures of genetic resources; **Plant Genetic Resources Newsletter**; v.108, p.1-16, 1996.

CARVALHO, C. de. et al. **Brazilian Fruit Yearbook 2010**. Santa Cruz do Sul: Ed. Gazeta Santa Cruz, 2010, 128 p.

CARVALHO, J.M. M. **BNB support for R&D in regional fruit growing**. Fortaleza: Banco do Nordeste do Brasil, 2009. v.4.

CAVALCANTI, J.J.V.; BARROS, L.M. Advances, challenges and new strategies for cashew genetic improvement in Brazil. In: SIMPÓSIO NORDESTINO DE GENÉTICA E MELHORAMENTO DE PLANTAS, 1, 2009, Fortaleza. **Proceedings...**Fortaleza: Embrapa Agroindústria Tropical, 2009. 210 p.

CAVALCANTI, J. J. V.; PAIVA, J. R. de; BARROS, L.de M.; CRISÓSTOMO, J. R. **Banco ativo** de **germoplasma** de **caju**: variabilidade, caracterização e utilização, 1997, 9p. Available at: <http://www.ceinfo.cnpat.embrapa.br/arquivos/artigo_2584.pdf>. Accessed on: 15 April 2011.

CAVALCANTI JÚNIOR, A.T.; CHAVES, J.C.M. **Production of cashew seedlings**. Fortaleza: Embrapa Agroindústria Tropical, 2001. 43p. (Embrapa Agroindústria Tropical. Documentos, 42).

CHITARRA, M. I. F. **Minimal processing of fruit and vegetables**. Lavras: UFLA/FAEPE, 2000. 113p.

COELHO, P. J. A.; BARROS, L. M. Collecting cashew germplasm. In: WALTER, B. M. T.; CAVALCANTI, T. B. (Technical Editors). **Fundamentals for collecting plant germplasm -** Brasília. Brazilian Agricultural Research Corporation - Embrapa Recursos Genéticos e Biotecnologia. 2005.

CORRÊA, M. P. F.; BUENO, D. M.; PARENTE, J. I. G.; PEREIRA FILHO, J. E.; ROSSETTI, A. G. Borbulhia: The economical graft for cashew trees. **Informativo**, Fortaleza, Embrapa Agroindústria Tropical, v.1, 1993.

CUQUMA, S. Plant Genetic Resources (PGR) and its importance in conservation, management & utilisation. **Technical Bulletin**, Nov. 2010. Available at: http://www.agriculture.org.fj/_resources/main/files/Issue%203%20Plant%20Genetic%20Re

so urces.pdf. Accessed on: 24 June 2011.

DESCHAMPS, C. **In vivo and in vitro vegetative propagation of Sarandi *(Sebastiana schottiana* Muell. Arg.), a riparian forest species**, 1993. 128 p. Dissertation (Master's Degree in Agronomy) - Lavras School of Agriculture, Lavras, 1993.

DUDLEY, E. D.; HOTCHKISS, J. H. Cysteine as an inhibitor of polyphenol oxidase. **Journal of Food Biochemistry**, [S.l.], v. 13, n. 1, p. 65-75, 1989.

FACHINELLO, J. C.; HOFFMANN, A.; NACHTIGAL, J.C.; KERSTEN, E.; FORTES, G. R. L. **Propagation of temperate climate fruit trees**. 2ª ed. Pelotas: Editora UFPel, 1995. 179p.

FOOD AND AGRICULTURE ORGANISATION OF THE UNITED NATIONS (FAO). **Plant genetic resources: their conservation** in situ **for human use**; FAO, Rome, Italy; 1989.

FRANKEL, O.H. AND BENNETT, E. **Genetic resources in plants: their exploration and conservation**; Blackwell, Oxford; 1970.

FRUIT: Brazilian fruit production up 19 per cent in eight years, says IBGE. **Brasil Atual**. Available at: <http://brasilatual.com.br/sistema/?p=3874>. Accessed on: 13 April 2011.

CEARÁ FOUNDATION FOR METEOROLOGY AND WATER RESOURCES (FUNCEME), 2011. **FUNCEME**, Fortaleza, 6 June 2011. Available at: < http://www.funceme.br/index.php/areas/rede-de-monitoramento/plataforma-de-coleta-de-data>. Accessed on: 6 June 2011.

FIGUEIREDO JUNIOR, H. S. Desafios para a cajucultura no Brasil: o comportamento da oferta e da demanda de castanha-de-caju. **Revista Econômica do Nordeste**, v. 37, n.4, p.550- 571, 2006.

GAZZOLA, R.; GAZZOLA, J.; COELHO, C. H. M.; SOUZA, G. S.; WANDER, A. E.; CABRAL, J. E. O. **Considerations on the cashew nut kernel - nutritional and market aspects** In: Congresso Brasileiro de economia e sociologia rural.
Londrina: Sober, p.65, 2007.

GOTO, R.; SANTOS, H.S.; CANIZARES, K.A.L. **Grafting in vegetables**. São Paulo: UNESP, 2003. 85p.

HAMMER, K. AND TEKLU, Y. Plant genetic resources: selected issues from genetic erosion to genetic engineering. **Journal of Agriculture and Rural Development in the Tropics and Subtropics**, s.l, v.109, n. 01, p.15-50, 2008.

HARTMANN, H. T.; KESTER, D. E.; DAVIES JR., F. T.; GENEVE, R. L. **Plant propagation:** principles and practices. New Jersey: Prentice-Hall, 880 p., 2002.

CEARÁ ECONOMIC RESEARCH AND STRATEGY INSTITUTE (IPECE). **Pacajus: basic municipal profile.** Fortaleza. 17 p., 2009.

KAHN, V. Effect of proteins, protein hydrolyzates and amino acids on o-dihydroxyphenolase activity of polyphenoloxidase of mushroom, avocado and banana. **Journal of Food Science**, Chicago, v. 50, p. 111-115, 1985.

KHOSLA, P.K.; SAREEN, T.S.; MEHRA, P.N. Cytological studies on hymalayan Anacardiaceae. **The Nucleous**, v.4, n.3, p.205-209, 1973.

MARIANTE, A. S.; SAMPAIO, M. J. A.; INGLIS, M. C. V. The State of BraziFsplant genetic resources. **Second national report: conservation and sustainable utilisation for food and agriculture**. Brasília: Embrapa Technological Information, 2009.

MELO, A. A. M.; VILAS-BOAS, E. V. de B. Inhibition of enzymatic browning of minimally processed 'Maçã' bananas. **Ciência e Tecnologia de Alimentos**, Campinas, v.26, n. 1, p. 110-115, 2006.

MITCHELL, J.D.; MORI, S.A. The cashew and its relatives (Anacardium: Anacardiaceae). **Memories on the New York botanical garden**, v.42, p.1-76, 1987.

MOLNAR-PEARL, I.; FRIEDMAN, M. Inhibition of browning by sulfur amino acids: 3. apples and potatoes. **Journal of Agricultural and Food Chemistry**, Easton, v.38, n. 8, p. 1652-1656, 1990.

OLIVEIRA, J.R.P.; SOARES FILHO, W. dos S. **Situation of the acerola crop in Brazil and Embrapa Mandioca e Fruticultura's actions in genetic resources and improvement.**
Rec. gen. e melhoramento de plantas para o Nordeste brasileiro, Petrolina, v.1, nov/1999. Available at: <http://www.cpatsa.embrapa.br/catalogo/livrorg/acerolabrasil.pdf>

PAIVA, J.R. de; CARDOSO, J.E.; BARROS, L.M.; CAVALCANTI, J.J.V.; ALENCAR, E.S. **Behaviour of dwarf cashew clones in the semi-arid region of the state of Piauí.** Fortaleza: Embrapa Agroindústria Tropical. 2001 (Embrapa Agroindústria Tropical. Comunicado Técnico, 63).

PAIVA, J.R. de; CRISÓSTOMO, J. R.; BARROS, L. M. **Recursos genéticos do cajueiro: coleta, conservação, caracterização e utilização**, Fortaleza. Brazilian Agricultural Research Corporation - EMBRAPA. 2003.

PEIL, R. M. Grafting in the production of vegetable seedlings. Grafting in the production of vegetable seedlings. **Ciência Rural**, Santa Maria, v.33, n.6, p.1169-1177, 2003.

PRODUCTION and cultivated area of early dwarf cashew increase in Ceará. **Frutal 2010.** Available at: <http://www.frutal.org.br/frutal2010/index.php?pg=show_noticia&cod=125>. Accessed on: 13 Apr. 2011.

RENDLE, A.B. **The Classification of Flowering Plants. Vol II - Dicotyledons** Cambridge University Press. 1938, 640p.

RIBEIRO, G.D.; COSTA, J.N.M.; VIEIRA, A.H.; SANTOS, M.R.A. **Grafting in fruit trees.** Porto Velho: Embrapa Rondônia, 2005 (Embrapa Rondônia. Recomendações Técnicas, 92).

RICHARD-FORGET, F. C.; GOUPY, P. M.; NICOLAS, J. J. Cysteine as an inhibitor of enzymatic browning: 2. Kinetic studies. **Journal of Agricultural and Food Chemistry**, Easton, v. 40, n. 11, p. 2108-2113, 1992.

ROCCULI, P.; GALLINDO, F.G.; MENDONZA, F.; WADSO, L.; ROMANI, S.; ROSA, M.D.; SJOHOLM, I. Effects of the application of anti-browning substances on the metabolic activity and sugar composition of fresh-cut potatoes. **Postharvest Biology and Technology**, Amsterdam, v. 43, n. 1, p. 151-157, 2007.

Brazilian Micro and Small Business Support Service - SEBRAE. **Agribusiness - Fruit Growing** (Intelligence Bulletin, 2015). Available at: <http://www.bibliotecas.sebrae.com.br/chronus/ARQUIVOS_CHRONUS/bds/bds.nsf/64ab 87 8c176e5103877bfd3f92a2a68f/$File/5791.pdf>. Accessed on: 15 Mar 2017.

SEIBERT, E.; CASALI, M. E.; LEÃO, M. L.; PEZZI, E.; CORRENT A. R.; BENDER, R. J. Cold damage and qualitative changes during refrigerated storage of peaches harvested at two stages of ripeness. **Bragantia**, Campinas, v. 67, n. 4, 2008.

SHAHIDI, F. Natural antioxidants: an overview. In: _ . **Natural antioxidants**: chemistry, health effects, and applications. Newfoundland: AOCS press, 1996. chap. 1, p. 1 - 11.

SIES, H.; STAHL, W. Vitamins E and C, P-carotene, and other carotenoids as antioxidants. **The American Journal of clinical nutrition**, Bethesda, v.62, n. 6, p. 1315-1321, 1995.

SILVA JUNIOR, J.F. da; BEZERRA, J.E.F.; LEDERMAN, I.E. **Genetic resources and improvement of native and exotic fruit trees in Pernambuco.** Rec. gen. e melhoramento de plantas para o Nordeste brasileiro, Petrolina, v.1, nov/1999. Available at: <http://www.cpatsa.embrapa.br:8080/catalogo//livrorg/fruteirasnativas.pdf>

SOBREIRA JÚNIOR, O. **Preservation and viability of cashew forks subjected to different types of packaging**. 2010. 48p. Monograph (Degree in Biological Sciences) - Federal University of Ceará, Ceará.

TREICHEL, M.; KIST, B.B.; SANTOS, C.E.; CARVALHO, C.; BELING, R.R. **Brazilian Fruit Yearbook 2016**. Santa Cruz do Sul: Ed. Gazeta Santa Cruz, 2016. 88 p.

WALTER, B. M. T.; CAVALCANTI, T. B. (Eds.). **Fundamentals for collecting plant germplasm**. Brasília. Brazilian Agricultural Research Corporation - Embrapa Recursos

Genéticos e Biotecnologia. 2005.

WILEY, R.C. **Minimally processed refrigerated fruits and vegetables**, London, CHAPMAN and HALL, 357 p., 1994.

ZACCARO, R. P.; DONADIO, L. C.; LEMOS, E. G. M. Micrografting in mango cultivars. **Revista brasileira de fruticultura**, Jaboticabal, v.28, n.3, 2006.

CHAPTER 8

APPENDIX - Preparation of preservative solutions.

All the products used to prepare the solutions (ascorbic acid, citric acid and cysteine) were weighed separately in their appropriate concentrations at the EMBRAPA - CNPAT Tissue Culture Laboratory in beakers, which were then immediately sealed with aluminium foil and plastic wrap and stored in a refrigerator at 8°C.

After around 72 hours, this material was taken in Styrofoam boxes to Pacajus-CE, where the forks were collected and grafted, where they were mixed with distilled water in a 2 litre beaker (FIGURE 12). In the case of preservative solution 1 (150mg/L ascorbic acid + 25mg/L citric acid + 2mg/L cysteine), the material was only mixed when the solution was prepared.

After the solutions had been prepared, the cuttings were separated into 3 groups of 21 cuttings, tied with elastic cord and immersed in the respective solutions for 2 minutes, after which they were either grafted or conditioned, according to the respective treatments.

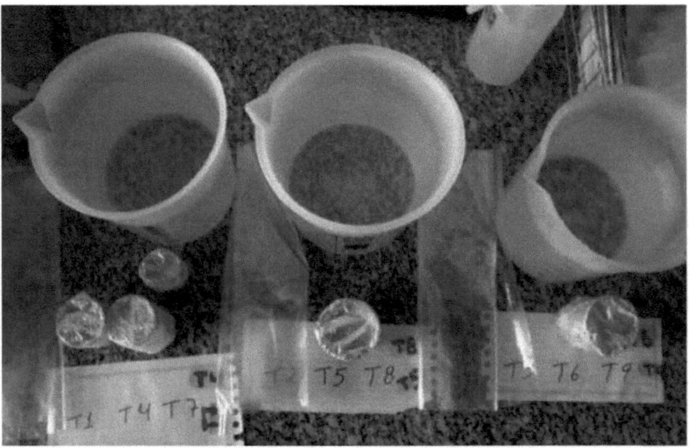

Figure 13 - Material used to prepare the solutions moments before immersing the cuttings.

ANNEX - Forecast and average yield of the Brazilian cashew nut crop for 2011

TABLE 1A - Forecast of the Brazilian cashew nut harvest for 2011 by federation units that stand out in national production.

Brazil *AND THE* **States**	**Safra**	**Planted area (ha)**	**Harvested area (ha)**	**Production (tonnes)**
Brazil	**2010**	770.285	754.863	105.885
	2011	767.408	756.060	295.688
Maranhão	**2010**	20.270	19.189	8.016
	2011	20.474	19.393	6.691

Piauí	**2010**	170.884	170.884	14.682
	2011	173.220	173.220	65.406
Ceará	**2011**	408.925	401.510	39.596
	2010	411.587	401.999	164.157
Rio Grande do Norte	**2010**	126.983	121.291	26.597
	2011	121.635	121.207	46.139
Paraíba	**2010**	7.065	7.032	2.735
	2011	6.979	6.974	2.304
Pernambuco	**2010**	10.007	9.109	8.819
	2011	7.247	7.247	5.475
Bahia	**2010**	26.151	25.848	5.440
	2011	26.266	26.020	5.516

Source: Modified based on data from the Systematic Survey of Agricultural Production - IBGE (2011)

TABLE 2A - Average cashew nut yield in Brazil between 2010 and 2011 for the month of February

Year		**Variation (%)**
2010	**2011**	
140	391	+179,29

Source: Modified based on data from the Systematic Survey of Agricultural Production - IBGE (2011).

Printed by Books on Demand GmbH, Norderstedt / Germany